Math Projects:

Organization
Implementation,
and Assessment

Katie DeMeulemeester

Dale Seymour Publications

With love to the people who inspired and supported the ideas in this book: the outstanding math faculty at Concord-Carlisle Regional High School.

Project Editor: Joan Gideon
Production Coordinator: Barbara Atmore
Art: Eulala Conner
Cover Design: David Woods

Published by Dale Seymour Publications,
an imprint of the Alternative Publishing Group
of Addison-Wesley Publishing Company.

Order Number DS21340
ISBN 0-86651-836-3

DALE
SEYMOUR
PUBLICATIONS
P.O. BOX 10888
PALO ALTO, CA 94303

1 2 3 4 5 6 7 8 9 10-MA-98 97 96 95 94

This book is printed on recycled paper.

Contents

Preface

Although mathematics is a broad field with widespread applications, the average secondary student is rarely exposed to all the facets of this subject. Reforms such as the NCTM *Curriculum and Evaluation Standards for School Mathematics* propose that mathematics teachers spend more classroom time with their students exploring and investigating applications of math to the real world. In addition, they recommend that students learn to value the cultural, historic, and scientific evolution of the subject.

An effective way to involve students in these pursuits is through personal investigation of a specific topic of their choice. Teachers who incorporate student projects into their curriculum often report positive changes in students' attitudes toward math class. Encouraging students to explore a topic in depth reveals the useful aspects of the subject, and provides students an opportunity to be an "expert" on at least one area of mathematics.

The authors of this publication and the companion volume, *Encyclopedia of Math Topics and References*, have personally experienced the advantages of student projects in their math classes at both the middle-school and high-school levels, but they acknowledge the difficulty of finding the time during the school year to come up with workable, innovative ideas and to design good assessment rubrics. Accordingly, this book contains 18 sample expectations sheets for projects assigned in classes ranging from middle-school math through calculus, as well as information and checklists to help you design, implement, and assess your own projects. We have attempted to include examples for all students, regardless of course or ability. The companion publication, *Encyclopedia of Math Topics and References* is a database containing hundreds of sample topics complete with library references; it will help students with their research.

We hope you and your students enjoy learning more about, and gain a deeper appreciation for, the beautiful and interesting field of mathematics.

Katie DeMeulemeester

How to Use This Book

This book is designed to help you come up with a project that suits *your* particular educational goals, whether you want to supplement or replace a curricular unit or allow students the freedom to pursue an outside interest. The following overview will help you maximize your results while minimizing your reading time.

- Part One contains specific examples for nine different presentation formats: videos, skits, photo albums, graphics portfolios, class books, games, posters, models, and business presentations. Each format has a detailed expectations form for a project assigned to middle-school students, as well as one for a project required in high-school courses. If you already know what type of project you want to assign, you may want to read just that section, then skip to Part Two.

- For teachers who have never assigned a project before, Part Two contains discussions of many features that you might want to consider (as well as pitfalls to avoid) when designing a project. Teachers who are more experienced with projects may only need to examine the Organizational Checklist at the end of the chapter (p. 43–45) to be sure they haven't overlooked anything.

- Part Three contains ideas on how to structure your classroom during the time that students are working on their projects, as well as sample handouts to streamline the process: letter to the family, behavior evaluation criteria, and forms for the students' partner and topic requests. In addition, for groups who wish to conduct their own research but who have not had training in statistical issues, there are sample interviews and surveys. The goal of this chapter is to supply you with many of the organizational tools you will need.

- Part Four deals primarily with classroom management issues during the presentations and assessment. The headers should help you skip to the sections that you find most relevant to the type of project you are assigning.

You are about to embark upon a bold new adventure—enjoy the trip!

Part One

Presentation
Formats

Presentation Formats

In the dark ages of mathematics education, *project* was often synonymous with *research paper*. While this is, at times, an appropriate format for displaying knowledge gained through the research process, there are other possibilities that are sometimes more successful in capturing students' interest and motivating them to excel. Several formats are described in this chapter, along with their applicability to different project types. You may find them useful in deciding which presentation vehicles you will allow your students to use on the particular assignment you are planning.

There are two sample expectation forms for each type of project; these sample assignments reflect a variety of topics and courses, at both the middle- and high-school level. Please feel free to use and modify any of the ideas for your classes. Teachers often find that it is easier to fine-tune ideas that have worked in another math instructor's classroom than it is to invent a project from scratch. Whatever format you decide on, you will want to spell out your particular requirements in your project expectations form.

Videos

Similar to a paper in intent, a videotape generally requires as much time to research and prepare. Its major strength is that the format allows students to show the phenomenon rather than describe it in words, and it gives them a chance to impress you with their creativity through inclusion of music, text, costumes, computer-generated graphics, etc. Required equipment includes a blank VHS tape, a video camera to record the action (handheld camcorders work well; a student merely needs an adapter cartridge to transform the tiny tape into a VHS-compatible format), and, preferably, access to two VCRs and another blank tape so students can edit their masterpieces before giving them to you for viewing. Check with your school or district about students borrowing this equipment if their families don't own it. Videos can also be showcased for future generations—students consider it the highest of all compliments to be asked for a copy of their tape for the teacher's video library.

One caution: some students get so caught up in the creative aspects of producing a video that they neglect the purpose of the assignment—the math they are supposed to be presenting. Be sure students are aware of all the requirements. Expectation forms like those on pages 5–6 let students know what is required. And don't forget to impose a maximum time limit—students having fun on tape have been known to go on for 45 minutes or more!

Sample Expectations Form

Video
Middle School Math Class

What

The math project you will be working on for the next week will be worth one test grade. Since we are studying area and perimeter, you will be responsible for finding an area of the school that needs carpeting or painting, determining the quantity of carpet or paint that would be necessary to do the job, and calculating the actual cost of purchasing the materials.

How

You will be working in a group of four (you will have a chance to request partners) to produce a videotape. We will have the school's videotape equipment in the classroom next week, and your team will sign up for two 10-minute time slots on two separate days—during the first you will gather measurements, then during the second you will explain on tape how to use this data to find the surface area and calculate the cost of the project. The final video should be 10 minutes long.

Grading Criteria

Be sure your video contains the following parts (You will be graded on the accuracy of your calculations.):
- Include a visual scan of the room in question. Show the team members carrying out their actual measurements. The floor space or walls you select must not be a perfect rectangle. [25%]
- Show a chart of the data for the various parts of the room, organized in a table. Explain how you used this data to calculate the surface area. [50%]
- Display several advertisements from the newspaper for prices of paint or carpet. Show how to use these prices to determine the smallest cost for completing the job. [25%]

Extra credit will be given to teams that show exceptional creativity in their original visual or auditory effects or to groups that attempt to find the area of an unusually difficult surface. Check with me if you think your team has an idea that may qualify you for bonus points.

Sample Expectations Form

What

The project on mathematical modeling you will be working on for the next three weeks will be worth one test grade. You must choose a periodic phenomenon encountered in everyday life, gather data generated by this phenomenon, use the data to create the graph of a sine or cosine waves, and derive its equation. Projects in the past have run the gamut from the ordinary (radio waves, sound or music waves, light waves, lasers, and pendulums) to the unusual (Ferris wheels, tsunamis, car engine piston displacements, lawn sprinklers, and better potato chip designs). I encourage you to select a topic that's unique or one that you find personally interesting.

How

You and your partner will produce a videotape. The final production should be edited for clarity, and should be between 10 and 20 minutes long.

Grading Criteria

- Introduction: Give some background or historical information on the topic you have chosen, and explain any technical terms or concepts. There should be clear indications that you have done some research for this section. [10 points]
- Data: Explain how you gathered and measured your data. Display it in tabular form. [10 points]
- Graph: Plot your data, clearly labeling both axes. Smoothly sketch the curve that best fits the data. [10 points]
- Equation Derivation: Use the data on your graph to derive the amplitude, phase shift, translation, and period of your sine or cosine wave. You must clearly explain on tape where each number comes from. [20 points]
- Application: Use your equation to predict future behavior of the phenomenon via two examples. For the first example you should provide a nontrivial value for the independent variable and solve for the dependent variable; for the second example please provide a nontrivial value for the dependent variable and, because the function is periodic, solve for at least three instances of the independent variable. [20 points]
- Error Analysis: Discuss some of the sources of error that may prevent your equation from accurately predicting the actual behavior of the phenomenon. [10 points]
- Technical Correctness: Your project must meet the specified requirements for length, be edited to remove extraneous scenes, and include a bibliography at the end. [10 points]
- Creativity and Effort: Did you put time and thought into selecting an original topic or coming up with an unusual point of view on a traditional idea? Was this project interesting to watch? [10 points]

Extra credit will be given to projects that evidence extreme effort through wonderfully creative visual or auditory effects, or by taking on an unusually difficult project that goes beyond the mathematical scope of what we have done in class.

Skits or Demonstrations

The poor man's video, a skit is often appropriate when your class doesn't have the time or equipment to produce a full-fledged videotape or when the topic being covered doesn't warrant full video treatment. Skits offer the same creative advantages mentioned above, but should be limited in length to about five minutes. Since the presentation is live, results will often be less polished and sillier than those you would get from a video, and you can't watch and grade them at home at your convenience, but it does allow for audience interaction. The only equipment necessary is whatever props the students choose to bring and a chalkboard for the mathematical part of the presentation. Requiring the class to take notes on the math will maintain a sense of purpose in addition to the fun atmosphere, and scheduling only one or two skits per day will prevent audience restlessness.

Sample Expectations Form

What

The math project you will be working on for the next two days will be worth one quiz grade. Since we are studying probability and combinatorics, I would like you to choose a short sample problem that someone with no mathematical background would find difficult to solve (like the handshake problem) and illustrate the solution.

How

You will be working with your cooperative-learning group of four to put on a short skit for the class. The length of your drama production should be no longer than eight minutes!

Grading Criteria

Be sure your skit contains the following parts:
- Start out by showing a situation where you might encounter this particular combinatorics scenario involving a very large number. For example, if you were explaining the handshake problem, you might start out by having your skit take place at summer camp, with an innumerate counselor suggesting that all one thousand campers take five minutes to get to know each other by shaking each others' hands, with no two pairs shaking hands at the same time. A good skit will have props and some sort of simple costumes. [25%]
- Use the "solve a simpler problem" strategy to demonstrate how to solve your particular problem for a smaller number. In the handshake problem, for instance, you could calculate the total number of handshakes that would take place if there were only two campers, then if there were three, then four. Once you have the answers to these simpler questions, show how you could use the math we have been studying to come up with the same solutions. Involve some members of the audience, if at all possible, and be sure to write your formulas and solutions on the board. [50%]
- Finish by using this mathematical technique to solve the original problem (in the example above, compute the total number of handshakes that would take place between all one thousand campers). [25%]

Extra credit will be given to teams that show exceptional creativity in their storyline or props or for groups that come up with an original problem that we haven't studied in class. Check with me if you think your team has an idea that may qualify you for bonus points.

Sample Expectations Forms

Skit
High School Geometry Class

What

The math project you will be working on for the next two days will be worth one quiz grade. Since we are studying the Pythagorean Theorem, I would like you to choose a short sample proof that you could use to convince someone who had been absent the day we proved the theorem in class. You may not use the same proof I showed you, nor may you use the same proof someone else in the class has claimed.

How

You may choose any two people in the class to work with. Your job is to create a short skit that could be used to present the Pythagorean Theorem in an entertaining way to a geometry class that has never seen a proof before. We will be going to the library tomorrow, where the three members of your team will do research into possible proofs that you could present; the librarian has put several books and copies of the *Mathematics Teacher* on reserve for your convenience. As soon as you have located a proof that you like, inform me so that I can be sure no other group will duplicate the one you have chosen. The remainder of the preparation time should be spent inventing a story line, determining what sort of visual aids you will need, and refining your presentation. The length of your skit should be no longer than 10 minutes!

Grading Criteria

- Did your skit have an interesting or humorous plot, complete with costumes or props? [20 points]
- Did everyone in the group take turns presenting parts of the proof? [20 points]
- Was your proof mathematically sound the way you presented it? [20 points]
- Were appropriate diagrams and visual aids used to make understanding the proof easier? [20 points]
- Was the rest of the class able to understand your proof? [20 points]
  (They will vote by secret ballot on whether your proof was intelligible or not.)

Extra credit will be given for teams that generate extreme audience interest or participation, use unique visual or auditory effects, or do some serious research and come up with an original proof that I haven't seen before! Check with me if you think your team has an idea that may qualify you for bonus points.

Photo Albums

A photo album is like a video broken down frame by frame, and it offers the unique opportunity for students to exhibit their mastery of both visual and written elements. The group takes photos of the phenomena. But, instead of a narrative voice-over, students write out their description of the mathematics in the picture. A picture camera may be easier to borrow and use than a video camera, and an album is certainly more time intensive, and therefore more polished and thoughtfully done, than a live skit, but the cost of purchasing and developing film can be prohibitive for some students. If you are lucky enough to have a photography class or club in your school that can help with the developing, a photo album can be inexpensively produced.

This large sunflower has 89 rows of seeds spiraling clockwise, and 55 rows spiraling counterclockwise.

Each pineapple scale is in three rows. It is in one of eight rows that spiral gradually, one of 13 rows spiraling at a medium slope and one of 21 steep rows.

Sample Expectations Form

Photo Album
Middle School Math Class

What

The math project you will be working on for the next week will be worth one test grade. Since we are studying Fibonacci numbers, I would like you to come up with several examples of Fibonacci numbers in nature.

How

You will work with a partner to produce a photo album of samples of Fibonacci numbers in real life. The easiest place to find appropriate images is to cut them out of magazines (nature magazines have great enlarged shots), although actual photos are also welcome.

Your album must contain at least five pictures, one of which will come from the work you do in art class this week. You have recently been working on still-life drawings. This week you will be sketching and shading pinecones. You will submit three sketches of the same pinecone from different angles. Be sure to position your pinecone properly before you start drawing it, so that your pictures can be used to illustrate a sequence of Fibonacci numbers. You should have at least four more pictures that show Fibonacci numbers in nature. Under each picture, write a paragraph explaining exactly where the Fibonacci numbers are. You may want to use different-colored highlighters to point out the relevant parts of the pictures.

Grading Criteria

This project will be graded by three teachers—math, art, and English.

- Your art teacher will assign $\frac{1}{3}$ of the grade for the quality of your sketches.
- Your English teacher will read the explanatory paragraphs and assign $\frac{1}{3}$ of the grade for technical correctness such as spelling, grammar, and style.
- I will assign the remaining $\frac{1}{3}$ of the grade for your ability to locate and describe the Fibonacci numbers in the photographs you choose.

Extra credit will be given to photo albums that contain more than the minimum number of examples or for examples that are unique (pictures of things no one else in the class thought to include).

Sample Expectations Form

Photo Album
High School Discrete Math Class

What

The math project you will be working on for the next week will be worth one test grade. Since we are studying codes, I would like you to come up with several examples of product codes that you run across in everyday life.

How

You will be allowed to choose one partner with whom to produce a photo album or scrapbook of product or bar codes that people in this country encounter everyday. There are several ways to find appropriate samples: leaf through magazines in search of those little subscription cards; sift through the daily mail for the post office's routing stamp; photocopy the ISBN number on the back of a library book; get an old imprint of your mother's MasterCard; or soak the UPC label off a can. Your album must contain five samples of different types of product codes, which you will analyze in detail.

Grading Criteria

Under each sample, describe the code as you would to someone who knows nothing at all about codes. Explain how the system works, then analyze the information contained in your specific sample. Each sample will be worth 20% of your project grade; points will be awarded for the correctness and thoroughness of your explanations and for the diversity of samples you choose to include (five subscription cards would not impress me). Your scrapbook should be about 10 pages long when you combine pictures and text.

Extra credit will be given for photo albums that contain more than the required number of different samples or for samples that are unique (types that no one else in the class thought to include). If your family has friends or relatives in the manufacturing business, you might want to contact them for information about their product—their code may well be different from the models we studied in class.

Graphics Portfolios

A graphics portfolio serves the same purpose as a photo album but requires more choice and creativity on the part of the student. Instead of merely capturing and analyzing images for their mathematical content to create a graphics portfolio, students use the math they have been studying to produce the pictures they are going to analyze. The images can be either computer-generated or hand-drawn, depending on the availability of high-tech equipment and the level of sophistication of the mathematical model the student is using. Since graphics portfolios will yield a wide variety of end results based on the initial assumptions of the students, it is a good choice when assigning a project of a dynamic nature; photo albums are more appropriate when studying static concepts.

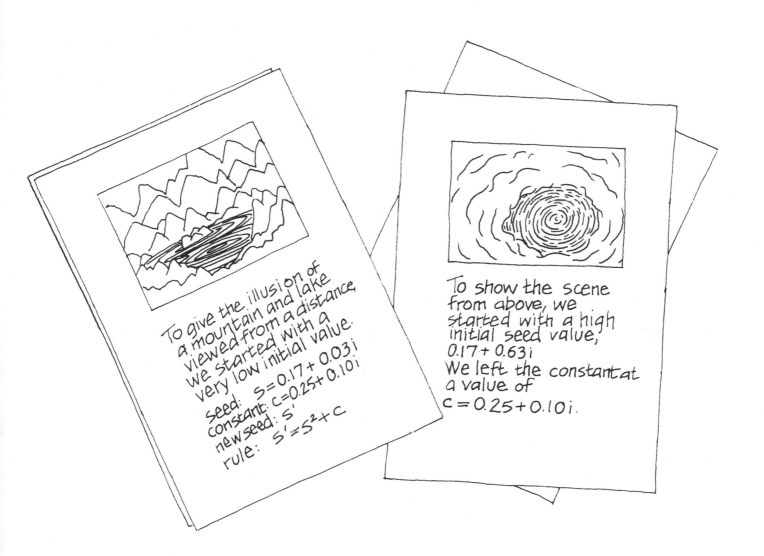

To give the illusion of a mountain and lake viewed from a distance, we started with a very low initial value.
seed: $S = 0.17 + 0.03i$
constant: $c = 0.25 + 0.10i$
new seed: S'
rule: $S' = S^2 + c$

To show the scene from above, we started with a high initial seed value,
$0.17 + 0.63i$
We left the constant at a value of
$c = 0.25 + 0.10i$.

Sample Expectations Form

Graphics Portfolio
Middle School Math Class

What

The math project you will be working on for the next three days will be worth one quiz grade. In class, we have been studying how to model real-life disasters with cellular automata techniques by selectively shading in the individual squares on graph paper. I would like you to come up with a "disaster" that you would be interested in modeling.

How

You will be working in teams of four to create a graphics portfolio; I will allow you to request a partner, then I will match two teams of two together to compose the final working group. You must decide on the specific type of disaster you want to study, whether it's an oil spill, a forest fire, or a biological crisis like a measles outbreak. Once your team has chosen a topic, you must each simulate the spread of the disaster under different initial conditions in order to gauge the varying impacts upon the final state of the system. For instance, in a team modeling a forest fire, Jamil might grid in the spread of the flames if there were no wind, Juana might grid in the spread of the flames if there was a 10-mph wind, and Daisuke and Lisa might shade in their grids to model 20- and 30-mph winds, respectively.

Grading Criteria

Each portfolio must contain
- one introductory paragraph, explaining the topic that was chosen [10%]
- four cellular automata graphs, with artist's name on top [10% each]
- one 200-word explanation attached to each graph, detailing the initial conditions of the model, the probabilities assigned to shading in the adjoining squares (the rule you used), and the method used to generate random numbers (such as dice, spinners, or programmable calculators), with author's name on top [10% each]
- one 100-word summary paragraph, drawing conclusions from the various final states of the systems [10%]

A grade will be assigned to each portfolio. The grade you personally receive will be the average of the group's portfolio grade and the percentage of points awarded to the specific graph you contributed. For instance, if the project receives a grade of 80%, but the graph and paragraph that you did got all 20 out of 20 possible points (100%), then your final grade would be the average of 80% and 100%—you would get a 90%.

Extra credit will be given to groups that go above and beyond the call of duty and include an additional section containing an actual case study at the end of their portfolio. For instance, the team that modeled forest fires might do some research into old newspapers to get a picture of a national park that has suffered a forest fire, and summarize the accompanying story to point out the initial conditions under which the fire started.

Sample Expectations Form

Graphics Portfolio
High School Algebra 2 Class

What

The math project you will be working on for the next two days will be worth one quiz grade. In class recently, we have been studying algebraic manipulations of imaginary numbers. You will work on the computer to see how imaginary numbers can be used to construct three-dimensional fractal landscapes.

How

You will be working in teams of four to create a graphics portfolio; I will allow you to request a partner, then I will match two teams of two together to compose the final working group. You must decide on the specific type of image that you want to create, whether it's a coastline, a mountain range, or decorative plant life, like ferns. Once your team has narrowed its focus, you will go into the Macintosh lab and each team member will run the fractals program starting with a slightly different seed value, in order to gauge the impact of small changes in initial conditions upon the final landscape that is created. The program works by taking an initial imaginary number, raising it to a power, adding a constant number, then iterating the process several hundred times. So if one partner starts by inputting $0.17 + 0.03i$, another may want to try $0.17 + 0.13i$, the third person may try $0.17 + 0.23i$, and the last member of the team $0.17 + 0.33i$. You are encouraged to experiment until you get a set of landscapes the entire group is happy with. Be sure to print a copy of the screen containing the data used to generate each landscape!

Grading Criteria

Each portfolio must contain
- four landscapes with their accompanying initial condition sheets, with the name of the team member who designed the specific graphic on top [5% each]
- one handwritten sheet attached to each graph, showing the algebraic steps involved in taking the initial seed value and cranking it through two complete iterations, to show how tedious it would be to create fractal images by hand [15% each]
- one 200-word summary page noting the differences in the various final states of the landscapes and suggesting a set of rules of thumb that a graphic designer might use in creating images with this program. Extrapolate from your results and hypothesize what changes in the picture would result from a new initial condition. [20%]

A grade will be assigned to each portfolio. The grade you personally receive will be the average of this portfolio grade and the percentage of points awarded to the specific image you contributed. For instance, if the project receives a grade of 80%, but the picture and paragraph that you did got all 20 out of 20 possible points (100%), then your final grade would be the average of 80% and 100%—you would get a 90%.

Extra credit will be given to groups that do some additional research into fractal images. Specifically, I want you to discover what basis the program uses to assign colors to the various parts of the landscapes. There is a great deal of recent literature on the subject; the school librarian will be able to help you conduct an on-line search for appropriate articles.

Class Books

A book which is produced by the entire class and given as a gift to a younger class is a wonderfully motivating way to complete a review unit. The maxim that you never understand anything as well as when you teach it is the driving force behind this assignment. Students are forced not only to dredge their brains for relevant ideas stored away last year, but also to actively consider the best way to present this knowledge to younger students who have never encountered it before. If the other math teacher is amenable, you can set aside a day or two where your students can go into another class, present copies of the book to the younger students, and work with them to master the subject matter. A particularly effective method is to place the novice students in ability-mixed, cooperative-learning groups of four, and match them with teams from your class, who will act as coaches. At the end of the review, a contest can be held, with prizes going to both the group that learned the most and the team that coached them. This is an opportunity for students to learn to view each other as math resources. It is also a time for the teachers involved to draw from the experiences of students as they discuss what types of explanation helped them to best understand the material. An additional benefit of the class book project is that it requires very little in the way of materials; paper, markers, and some sort of binder are really all that are necessary.

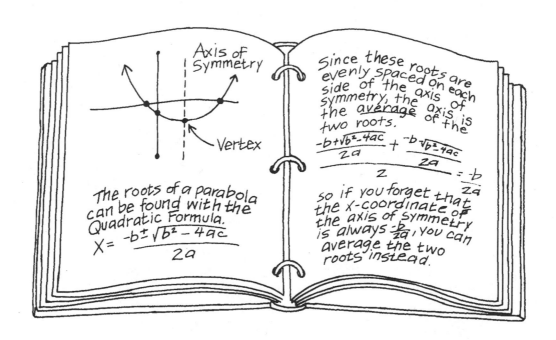

Sample Expectations Form

Class Book
Middle School Math Class

What

The math project you will be working on for the next three days will be worth one test grade. Last year in the fifth grade you learned the rules for adding, subtracting, multiplying, and dividing fractions. Before we can move on to higher-level concepts this fall, I need to see how many of those skills you remember.

How

You will be working in your cooperative-learning groups of four to create a book that will be given as a gift to clarify the mysteries of fractions for this year's fifth graders. After discussing the rules with your teammates, one person should be assigned to write each section. Later, all of you will chip in together to write a sample final exam at the end of the book.

Grading Criteria

Each book must contain one section for each of the following concepts: adding fractions, subtracting fractions, multiplying fractions, and dividing fractions. Each section is worth 20 points and will be graded on the accuracy of the following:

- Is there a clear explanation of the mathematical rule involved, complete with an example worked out step by step? [10 points]
- Is the section neatly done, and does it contain useful graphics that might help the students understand the concepts more thoroughly? (For example, show $\frac{2}{3}$ of a pizza minus $\frac{1}{2}$ of a pizza.) [5 points]
- Is there a set of five practice exercises, complete with an answer key worked out on another page? [5 points]

The final exam at the back of your book should be 20 questions long, and should contain a variety of questions from each of the four sections (like the practice exercises you wrote, but with different numbers), all mixed up in no particular order. Attach a removable answer key to this test to the back of your book. [20 points]

A grade will be assigned to each book. Then, after all the books are collected, you will individually take a review exam I have written. The grade you receive for this project will be the average of the grade your team receives on the book and the grade you score on this exam. For instance, if the book receives a grade of 100%, but you only score 80% on the review exam, then your final grade would be the average of 80% and 100%—you would get 90%. You should work with your teammates to be sure that everyone in your group can successfully solve all the problems on the final exam you write.

Extra credit will be given if every member of your group scores above 85% on my review exam.

Sample Expectations Form

Class Book
High School Pre-Calculus Class

What

The math project you will be working on for the next three days will be worth one test grade. Last year you studied conic sections: circles, parabolas, hyperbolas, and ellipses. Before we can move on to higher-level concepts, I need to see how much you remember.

How

You will be working in groups of four to create a book that will be given to this year's second-year algebra class to clarify the mysteries of conic sections. After discussing the concepts with your teammates, assign one person be to write each section. Later, all of you will chip in together to write a sample final exam at the end of the book.

Grading Criteria

Each book must contain one section for each of the following concepts: circles, parabolas, hyperbolas, and ellipses. Each section is worth 20 points and will be graded on how well you provide or explain the following:

circles

- formal definition [5 points]
- $x^2 + y^2 = r^2$ and $(x - h)^2 + (y - k)^2 = r^2$ forms [5 points]
- completing the square to get to center-radius form from $x^2 + y^2 + ax + by + c = 0$ [5 points]
- five practice exercises complete with an answer key worked out on another page [5 points]

parabolas

- formal definition (terms: *focus, directrix*) [5 points]
- finding the vertex, roots, and *y*-intercept from the form $y = ax^2 + bx + c$ [5 points]
- finding the vertex, roots, and *y*-intercept from the form $y = a(x - h)^2 + k$ [5 points]
- five practice exercises complete with an answer key worked out on another page [5 points]

hyperbolas

- formal definition [5 points]
- how to put in standard form and graph, and how to tell if it will open up or down or side-to-side [5 points]
- terms: *foci, vertices, transverse axis* [5 points]
- five practice exercises complete with an answer key worked out on another page [5 points]

ellipses

- formal definition [5 points]
- how to put in standard form and graph [5 points]
- terms: *foci, eccentricity, major axis, minor axis* [5 points]
- inclusion of a set of five practice exercises, complete with an answer key worked out on another page [5 points]

(continued on page 19)

Sample Expectations Form

Class Book
High School Pre-Calculus Class

Working together, write a final exam for the back of your book that is 20 questions long, containing a variety of questions similar to those from each of the four sections (though slightly different than the practice exercises you wrote), all mixed up in no particular order. Attach a removable answer key to this test to the back of your book.
[20 points]

A grade will be assigned to each book. Then, after all the books are collected, you will individually take a review exam that I have written. The grade you receive for this project will be the average of the grade your team receives on the book and the grade you score on this exam. For instance, if the book receives a grade of 100%, but you only score an 80% on the review exam, then your final grade would be the average of 80% and 100%—you would get 90%. As you can see, it is crucial that you can successfully solve all the problems your group puts on its final exam.

Extra credit will be given if every member of your group scores above 85% on my review exam. Please be sure you help each other review these skills.

Games

Another useful format for review, whether of last year's material or of material they have just learned, is to have the students create board games. This is one of the most popular projects because—quite simply—students love to play games. Whether it's a basic math spin-off of a quiz show or a more complex version that involves overhauling an old board game, students approach this activity with enthusiasm and many original ideas. The supplies needed will vary from class to class; the easiest way to come up with spinners, dice, and boards is to ask students to bring in old family games that can be cannibalized. Be sure to keep plenty of scissors, white-out, glue, and tape on hand for last-minute repairs and design changes. Warn the class that fragile games won't last long—they should use thick cardboard for their question cards and keep moving parts to a minimum. An important part of the assessment of the completed project is to have the students rotate in groups through at least two games apiece, playing each for about 20 minutes. Ask each group that plays a game to provide feedback on their enjoyment of the game and its usefulness as a review mechanism. The students will feel invested and involved, and you will gain a wonderful supply of engaging review materials that can be used with future classes.

Sample Expectations Form

Game
Middle School Math Class

What

The math review project you will be working on for the next two class periods will be worth two quiz grades. Your assignment is to design and create a working board game that another group can use to review for the test on scientific notation and decimals.

How

Our junk trunk contains donated old games that can be recycled into any form you desire. You will be working with your assigned cooperative-learning group of three people to choose a recycled game as a base for inventing your own Decimal Derby game.

Schedule

- Today: pick your game and decide on a master plan for the new creation. You should design a game that will yield a winner after about 20 minutes of play. Movement around the board should be controlled by some sort of demonstrated knowledge of scientific notation and the rules for addition, subtraction, multiplication, and division of decimals.

- Tonight: for homework everyone on your team should work on the portion of the game that has been assigned to them. Although tasks will differ from group to group, please be sure the following things get done:

  The person in charge of *rules* should type a summary of the object of the game, a list of preparation chores (like distributing start-up supplies or selecting who goes first), and a thorough description of how the game is played, specifying any time limits or problem situations that you can foresee. Don't forget to tell how to win the game.

  The person in charge of *artwork* is responsible for redecorating the cover of the box and redesigning the board spaces to reflect the new purpose of the game.

  The person in charge of *equipment* should create any question cards (including answers) and markers necessary to play the game and should also stock the game box with all essential materials like pencils, scratch paper, dice, spinners, or timers—whatever your group decides is appropriate.

- Tomorrow: your team will have half an hour to put the game together and take it for a trial run, ironing out any difficulties that arise along the way. Be sure you have timed the game so that someone wins within 15 to 20 minutes. If the game is clearly too long, you may have to modify the rules somewhat to make it shorter.

- Tomorrow night: you will write an essay detailing the tasks and ideas that you contributed to the creation of this game, and suggesting the grade you think the project should earn. It is very important that you take the time to do this carefully, since I will refer to this sheet when I am assigning a grade to your component of the project. I'd also like you to comment briefly on your opinions of the effort your teammates put into this project; your responses will be kept strictly confidential.

- Next week: two randomly selected teams will play your game and rate it.

(continued on page 22)

Game
Middle School Math Class

Grading Criteria

- Part of your grade will be based on the players' review of how well the game meets three important criteria. [50%]

  1. Could you figure out how to play the game by reading the rules or did you need to ask for instructions from the creators?

  2. Did you need to know information about scientific notation and decimals to be successful at this game? Was it helpful as a review tool?

  3. Was this game well designed and fun to play? Are there any improvements you might suggest?

- A group grade will be based on my personal assessment of the three questions above. [25%]

- A grade given to you alone will be based on my observations of the work you did in class, my assessment of the quality and effort put into the component that you contributed to the group project, and your responses to the homework essay. [25%]

Extra credit will be given for games that are of extremely high caliber or that evidence an unusual level of effort and creativity.

Sample Expectations Form

Game
High School Calculus Class

What

The math review project you will be working on for the next 30 minutes will be worth one quiz grade. Since we have finished the section on limits, I want you to focus on the distinct types that some of you had trouble with earlier this week, so you will know which sections you need to spend additional time reviewing before the test.

How

You will be working in your assigned cooperative learning groups of four to create sample Advanced Placement questions for the game Trivial Pursuit®. Since there are six groups in the class, each group will be assigned to write the questions for a specific color pie-piece, and the categories are

- pink—limits involving the natural logarithm or its base, e
- yellow—limits involving trig functions
- orange—limits involving radical expressions
- brown—limits in the form zero divided by zero
- green—limits in the form infinity divided by infinity
- blue—limits involving asymptotes or piecewise functions

Your group should come up with eight different questions for the category you are assigned, then make six copies of each question (48 cards in all). You may use your books or notes to generate ideas for the questions you want to ask, but the questions you write must use different numbers from the example in the text. Don't forget to work out the solutions and put the answers on the backs of the cards. Then tomorrow a set of each team's questions will be distributed to each of the six game boards, and each group member will go

to a game table to compete against three other people of similar ability from different groups, using the questions the class has generated.

Grading Criteria

At the end of the period, everyone in the class will rate each category of question on a scale of 3 (the group did an excellent job), 2 (the group did an adequate job) or 1 (the group did a terrible job), addressing the following issues:

- The questions were similar in level of difficulty to the ones worked out in class and assigned on the homework.
- There were no mistakes in the answers on the backs of the cards.

Points will be tallied and converted to a percentage score, which is the grade you will receive. In addition, the winners at each ability-based game table will have an extra 5 points added on to their scores.

Posters or Charts

Next to traditional research papers, the presentation vehicle most popular with math teachers is the poster, or chart. And for good reason—a poster is an ideal way to show distilled knowledge, attractively spruced up with related graphics and catchy titles. Posters can be used with any reasonably short topic, ranging from the concepts individual students have researched to the results of surveys the entire class has conducted. Students will need markers, magazines, glue, tape, and large pieces of cardboard. The primary advantage of posters is that they are easy to make, appeal to the artistic qualities in students, and can usually be completed in about two days. Doing the research or gathering the data for the poster may take a week or more. Posters and charts make excellent beginning-of-the-year projects, both because they acclimate the students to your alternative assessment techniques in a nonthreatening way and because they brighten up your room before back-to-school night.

A fractal called a Sierpiński Gasket can be formed from Pascal's triangle. Put each number into its own triangle. Color in every triangle that contains an even number—a pattern will be created.

"Pascal's Triangle" Misnamed

There is historical evidence that the Chinese discovered this array of numbers at least 300 years before Blaise Pascal was even born!

How to Get the Triangle

Start with ones on the outside of the triangle. To fill in a number within the triangle, add the two numbers immediately above it.

Repeat to get the remaining lines.

Sample Expectations Form

What

The math project you will be working on for the next two days will be worth one quiz grade.

The assignment is to research some aspect of Pascal's triangle, either its history or one of the many applications of the triangle to different mathematical fields.

How

You will be working with a partner of your own choosing to research the triangle and to design a poster that communicates your findings. The school librarian has on reserve a stack of *Mathematics Teacher* magazines that contain relevant articles, and should you require greater depth, the town library has the equipment necessary for you to conduct an on-line search. Once you have decided on a specific topic, clear it with me so that I can make sure someone else isn't already working on your particular area of research. In addition to the poster, each team of two will give a three- to five-minute oral presentation of their findings to the class sometime over the course of the next month. Sign up for your preferred presentation date on the calendar posted on the wall next to the door.

Grading Criteria

- Research: Have you been able to come up with unique applications or facts that we did not discuss in class? Did you include a bibliography of your sources of information? [30%]
- Artistic Appeal: Does your poster neatly and attractively present the information that you gleaned from your research? Have you included pictures along with your explanatory text? [40%]
- Oral Presentation: Do you and your partner take turns explaining the interesting facts and applications you discovered? Are your explanations clear, accurate, and concise? [30%]

Sample Expectations Form

Poster
High School Probability and Statistics Class

What

The math project you will be working on for the next two days is titled "Gambling: a great way to lose money" and will be worth one quiz grade. In class we did an extremely detailed analysis of the odds involved in the casino game craps. You learned why this game always favors the house and how much money the casino can expect to make every time someone plays. Your assignment is to research any other game of chance and summarize your findings on a poster. One caution: it is easier to calculate the probabilities involved in single-round games (like spinning a roulette wheel or buying a lottery ticket) than it is to analyze continuous draw card games (like blackjack). Please keep this in mind when selecting your game.

How

You will be allowed to work with a partner of your own choosing. The school librarian has on reserve a stack of books and magazines that contain relevant articles, and should you require greater depth, the town library has the equipment necessary for you to conduct an on-line search.

Once you have decided on a specific topic, clear it with me so that I can make sure no one else is working on the same game. In addition to the poster, each team will give a five-minute oral presentation of their findings at the start of class sometime during the next month. Sign up for your preferred presentation date on the calendar.

Grading Criteria
- Game Research [40%]
  Your poster should contain a typed sheet that details the following information: origins of the game, rules for playing it, payoffs and odds,

how much money you would expect to lose if you played the game 10,000 times and always chose the bet most favorable to you, how much money you would expect to lose if you played the game 10,000 times and always chose the bet most favorable to the house, a bibliography of your sources.
- Societal Impact Research [20%]
  Attach to your poster a newspaper or magazine article telling the story of someone who has ruined his or her life because of compulsive gambling. Include at the bottom the name and date of your source.
- Artistic Appeal [20%]
  Your poster should neatly and attractively present the information that you gleaned from your research. Include pictures along with your explanatory text, and summarize any critical expected-loss information in tabular form.
- Oral Presentation [20%]
  You and your partner should take turns explaining the rules and probabilities you researched. How does the expected loss of this game compare to others that the casinos offer? Your explanations should be clear, accurate, and concise.

Models

Closely related to posters, models are a three-dimensional way to illustrate a short topic. They are particularly relevant in geometry, where the results of many interesting theorems can be constructed for a modest investment in string, straws, toothpicks, or Tinkertoys®. Models also provide a natural way to incorporate math into other disciplines. A trigonometry or math analysis class can get a healthy dose of physics and engineering by building models of structures like the poorly designed Tacoma Narrows Bridge, and can subsequently use their models to study the concept of resonance to understand the inevitability of the bridge's collapse. Like posters, models do not require a large investment of money or class time. They are, however, usually quite fragile and need a display space that is out of the way of well-intentioned but grasping hands.

Sample Expectations Form

Model
Middle School Math Class

What

You will spend the rest of this class period working on a math project, which will be worth one quiz grade. You will be building several gumdrop/toothpick models to provide support for your conjecture linking the number of edges, faces, and vertices of any polyhedron.

How

Your cooperative-learning group of four will be subdivided into two teams of two. One team will create a tetrahedron and an octahedron; the other will build a pentahedron and a hexahedron. Group members should then share their models to brainstorm about possible relations between the number of edges, faces, and vertices of each polyhedron. Individuals in the group of four are responsible for compiling their own charts showing the data for each figure, and then for writing a summary paragraph showing how the numbers support the team's hypothesis.

Grading Criteria

- Models: Are the four models properly built? [50%]

- Data: Is the data in the chart accurate? [25%]

- Conjecture: Is the conclusion reached supported by the data? [25%]

Sample Expectations Form

Model
High School Algebra 1 Class

What

You will spend today and tomorrow working on a math project, which will be worth one quiz grade. Yesterday in class we talked about the slope of lines. Now I want you to design and build a device that can be used to find the slope of a given line segment.

How

You and your cooperative learning partner will each build a SlopeFinder. One of you will create a device that measures positive slopes; the other will build a model that measures negative slopes. You may be more successful if the two of you brainstorm a general idea of what this device will look like and then modify it to calculate the slope of the type of line you were assigned.

Use your creativity. You can build the SlopeFinders out of any materials you like. The only restrictions are

- They should work on any line segment of length 10 cm or greater.
- They should be durable enough that they won't break or fall apart if someone other than yourself uses them.

You will have the last 20 minutes of class today to come up with a design plan for your SlopeFinder and to decide what materials you will need to build it. I expect you to work on your model tonight for homework and will give you the first 15 minutes of class tomorrow to fine tune it with your partner.

Each individual is also responsible for writing a summary paragraph explaining how to use both devices and describing to me why they give you the same solution you would get if you used the algebraic formula

$$m = \frac{y_2 - y_1}{x_2 - x_1}$$

In other words, how does your physical model relate to the algebra we studied?

Grading Criteria

- Models [50%]

  The two models are properly built and perform the functions they are supposed to.

- Explanation [50%]

  Your paragraph clearly explains how to use the SlopeFinders and correctly describes the connection with the algebraic solution.

These devices will be displayed at the math department meeting next week, and faculty members will be asked to vote on which product they like best. The top three winning design teams will receive extra credit points. Note: An excellent SlopeFinder might go beyond the constraints of this assignment and also handle vertical and horizontal lines.

Business Presentations

The final method of displaying gained knowledge is the formal business presentation. This vehicle is most appropriate for long, involved projects where the students have had to gather and analyze data. Since one of our goals as teachers is to prepare our students for the real world, this type of project—though time intensive—is one of the most valuable you can assign. Ask students to determine a real need within their school or community. Using proper information-gathering techniques, they work in groups to write and administer a survey that measures the need and solicits opinions of possible solutions. Then, perhaps with the aid of business or graphing software, the team can transform their raw data into readable bar graphs and pie charts. They can put these graphics on chart or transparencies and prepare a five-minute oral presentation to familiarize the class with their findings.

This type of project lends itself naturally to outside speakers from the community—you may want to have some businesspeople come in to model presentation skills for the class and to help the students learn to use the software packages. Some volunteers have even been known to "adopt" a group of students, bringing them to their workplaces so that the students could view actual presentations or use the latest technology. Similarly, you should be willing to reward outstanding student presentations—those that come up with viable solutions to an actual need—by arranging to have those students give their presentation to the appropriate people, whether it be your school's faculty or the city council. This is an incredible opportunity for your classes to realize their own importance in the civic process, and to appreciate the mathematics that allows them to intelligently assess a situation.

Sample Expectations Form

Business Presentation
Middle School Math Class

What

Because next week is Drug and Alcohol Awareness Week, you will spend this week working on a math project—a presentation based on relevant drug statistics that you have gathered and researched—which will be worth one test grade.

How

I'd like you to request the names of three other people that you think would work well with you to turn out a quality product. Your grade on this project will reflect not only the quality of your data-gathering methods and of your statistical analysis, but also your team's skills in using the computer to produce graphics and in giving an effective oral presentation. For this reason, it is important that you choose partners whose abilities and talents cover these major areas.

Topic

Please pick a drug, either legal (like caffeine, alcohol, nicotine) or illegal (such as marijuana, cocaine, LSD) that you and your partners are interested in conducting a use survey on. Depending on the drug you pick, your target survey audience may range from middle-school or high-school students to adults.

Project Components and Grading Criteria

- Research: Go to the library and find some background information on your drug. Research questions such as, how is it produced or grown? What are its effects on the human body (both short- and long- term)? How addictive is it? This information should be distilled into a typed, one-page synopsis. Be sure to include a bibliography of your sources. You will be graded on the thoroughness of your research and the grammatical correctness of your paper. [25%]

- Survey: Using the questioning techniques we studied in class, design a survey to gather information about the frequency of use, in a local target population, of the drug you are studying. Choose your demographic audience carefully, keeping in mind that your team will have to administer this survey to at least 60 randomly selected people for the results to have any statistical significance. In past years some groups have had older brothers and sisters administer the survey in their high-school or college classes, and some have had another family member give it to colleagues at work or to neighbors. Please write a one-paragraph description of how you conducted the polling and staple it to a copy of your survey. You will be assessed on the quality of your survey (does it meet the criteria we outlined in class?) and on whether proper randomizing techniques were used when it was administered. [25%]

(continued on page 32)

Business Presentation
Middle School Math Class

- Statistical Analysis: Use the data from your survey to make some predictions about the use of your drug within the local population. The data should be cut several different ways, depending upon the questions you asked. It is often informative to break usage down by age or sex (point out any trends that you notice). You should prepare eight overhead transparencies that show some or all of the following (relevance will vary depending upon your survey): data distribution—frequency tables, mean, median, mode, standard deviation; graphs—scatterplots, pie charts, bar graphs, box-and-whisker plots.

  Use the computer to create graphs that will make an impact upon the audience. Even though we spent two days in the computer lab last week, I know some of you are still not comfortable working with the business software. Remember, the computer specialist has set aside periods this week to be available to help you; your team merely needs to ask for assistance. Your grade will be based upon the mathematical soundness of your data analysis, as well as on your demonstrated ability to use software packages to produce quality graphics from your raw data. [25%]

- Formal Presentation: There is a sign-up sheet on my desk; two groups per day need to present on Monday, Tuesday, Wednesday, and Thursday next week, but you are welcome to choose the day that is most convenient for you. Remember to dress professionally and to rehearse the night before so that you are not overly dependent upon your note cards. Everyone on your team must speak (at least briefly) as you present your research and survey findings to the class. Please limit your presentation to 15 minutes. If you finish early, you are welcome to solicit questions from the audience. Your grade will reflect how well-prepared and informative your talk is. [25%]

- Bonus Points: The team that receives the highest score will be invited to give their presentation at the faculty meeting on Friday of next week. This is your big chance to instruct your teachers. (No pop quizzes, please.) An additional 10 points will be added to your score for the extra work involved in making the presentation to the faculty.

Sample Expectations Form

What

Many of you have expressed dissatisfaction with the fact that our town does not have a teen center where teenagers can go to hang out after school. To explore the possibilities of a teen center, you will spend the next month preparing a presentation of a proposed teen center based on your research and statistics you have gathered. This math project will be the basis for your quarter grade. After all the presentations have been given, the class will vote on the components that they like best, and a group will be elected to combine all these elements into one final presentation to present to the town officials at next month's city council meeting.

How

Request your choice of three people that you think would work well with you to turn out a quality product. Your grade on this project will reflect not only the quality of your data-gathering methods and of your statistical analysis, but also your team's skill in using the computer to produce graphics and in giving an effective oral presentation. For this reason, it is important that you choose partners whose abilities and talents cover these major areas.

Project Components and Grading Criteria

- Week One: The Survey

  It is important that your proposed teen center reflect the needs and desires of as many local teenagers as possible. Using the questioning techniques we studied in class, design a survey to gather relevant information such as: What areas should there be (TV room, kitchen, study room, video-game arcade, dance floor)? Where should the center be located? How should it be staffed (academic support, counselors)? What hours should it be open? What sports equipment should it maintain? The more thorough your survey is, the better your design will be. Your team will have to administer this survey to as many randomly selected teenagers as possible (at least 100) for the results to have any statistical significance. Please write a one-paragraph description of how you conducted the polling and staple it to a copy of your survey. You will be assessed on the quality of your survey (does it meet the criteria we outlined in class?) and on whether proper randomizing techniques were used when it was administered. [25%]

- Week Two: Statistical Analysis and Computer Graphics

  Use the data from your survey to determine the needs of the people who will be using the center. Use the business software you learned to use last week to show the frequency of the audience responses, both in tabular form and with pie charts and histograms. You should also break the responses down by age and by sex to see if there are subgroups within the population who have different needs. For the numerical answers to survey questions (strongly agree—strongly disagree), you should calculate the mean, median, mode, and standard deviation, and portray these results with box and whisker plots. Your grade on this section will be based upon the mathematical soundness of your data analysis as well as on your demonstrated ability to use software packages to produce quality graphics from your raw data. [25%]

- Week Three: Research and Design

  Once you have the results of your statistical analysis, it's time to design the actual teen

Business Presentation
High School Consumer Math Class

center. The first thing you need is a site. Based on your respondents' preferences, choose an area of town and find an appropriate building the town could use to house the center. Research the cost of purchasing or renting this site, or of buying land and building the center from scratch. Also, arrange in tabular form the cost of all the staff salaries and equipment that would be necessary to fund the center on a year-round basis. I expect to see documentation for each expenditure—if you buy a TV, I want to see an ad listing the price; if you say the cost of a part-time counselor is $15,000 per year, I want to know who you called to get that figure. Your group may want to visit the teen centers in neighboring towns to see how they are run and to get a copy of their budgets for comparative purposes. If your proposal is to have any hope at all of becoming reality, the figures you use need to be accurate. Finally, draw a scale model floorplan, using architectural software (I will show you how to use it in class), showing the location of the rooms and equipment. You will be assessed on the thoroughness of your budget, the strength of your documentation, and the quality of your floor plan. [25%]

- Week Four: Formal Presentation

There is a sign-up sheet on my desk; one group per day needs to present on each day of that last week, but you are welcome to choose the day that is most convenient for you. Remember to dress professionally and to rehearse the night before so that you are not overly dependent upon your note cards. Everyone on your team must speak (at least briefly) as you present your survey or research findings and your design proposal to the class. Please limit your presentation to 25 minutes, and be prepared to

encounter feedback from your classmates, who may either support or disagree with your statements (remember, they have been researching the same issues). You will be graded on how well-prepared your presentation is, how appropriate your use is of computer-generated slides, and how well you respond to audience criticisms of your data. (Can you support your assertions?) [25%]

Everyone is expected to take notes on components of other groups' proposals that they strongly like or dislike, so that at the end of the week the class can vote on what to put in the final proposal that goes to the city council. Groups that authored the relevant pieces will elect a representative that will work on polishing the proposal and presenting it at the next town meeting.

Part Two

Organization

Organization

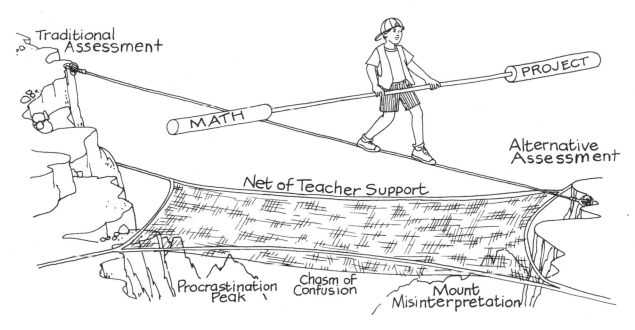

While the expectation forms in Part One illustrated some different formats, chances are that none of the sample projects exactly met your specific needs. This chapter will provide you with an overview of issues to consider when designing your own project. Keep in mind that a project can contain as many or as few of these items as you feel are appropriate. After all, you are the teacher. You know best the abilities and background of your students. The end of this chapter contains a checklist to help you organize the key points you want to include in your project design.

Organization is the key to having a successful and enjoyable experience with projects. It is crucial to have completely mapped out ahead of time the goals, time frame, and expectations for the assignment. Keep in mind that a well-designed project should not increase your workload; it should replace some of the activities you haven't been totally satisfied with in the past. Here are some issues to consider while planning that will help teacher and student alike have a positive experience.

Topic

The place to begin is with the subject matter. Pick a topic or unit from the curriculum that you think lends itself to student exploration, and decide exactly which sections the project will cover. It will help your students if you try to be as specific as you can: "modeling periodic motion using sine and cosine waves" is better than "trig equations"; similarly, "designing a roller coaster using force vectors" gives them a more accurate focus than "the vector project." If you prefer to allow students to choose their own topic, you may want to offer a list of sample titles, which indicates the degree of specificity that you desire. Working with a teacher from another subject is an excellent way to generate student interest in projects—see the section on coplanning for ideas on selecting cross-curricular topics.

Presentation Format

Envision the type of product you would like back. Do you want videos, term papers, class presentations, or another presentation form? There is room for flexibility, and some students will enjoy designing their own format. But many others will want guidance on the types of projects you consider acceptable.

Purpose

Cognitive Goals

What learning objectives do you hope to accomplish by assigning this project? If you are covering a unit on statistics, for instance, you may want students to be able to design a survey to generate data, analyze the data for the line of best fit, and present their results graphically in several different forms. Be sure your project outline clearly provides for these objectives, or else create supplementary assignments (worksheets, lectures) that cover the ones left out.

Background

What skills or knowledge must the students possess before they begin? Is it safe to assume that they have mastered these concepts, or will you have to provide a review using worksheets or lectures before beginning the project? A three-week unit on probability, for example, could quickly be derailed if your sixth grade class has forgotten how to manipulate fractions.

Social Goals

Are there any social skills that you want the students to practice while working on their projects? Remember that students often need clear instruction regarding acceptable behavior in cooperative learning situations. Therefore it might be worthwhile to specify the desired behavior (pick only one or two things to work on at a time) and plan a system that provides feedback and evaluation of the behavior in question. Assigning roles (reader, recorder, encourager), modeling positive feedback (I-statements), and creating grade interdependency are all proven techniques that encourage students to work at building healthy, interactive relationships.

Time Frame

As you organize, think about both your time and your students' time.

Students' Needs

How much time will your class need to do a thorough job on the project, yet not get tired of it? You will have to divide this time into the number of class periods you are willing to give up and the amount of time that you can realistically expect them to devote to the assignment at home. Remember that students working outside of your supervision will tend to procrastinate. It will help if these tasks are broken into smaller pieces with staggered due dates. Now is the time to plan your policy on giving extensions and assigning penalties for late projects.

Your Needs

Guaranteed: students want immediate feedback, but if you collect and try to grade 75 projects at once, you will never assign another one! How can you prevent this from happening to you? If it is a unit project, where all students are covering the same curricular material, it may be possible to set aside a series of due dates over the course of a month and let teams choose the one that works best for them. This way you will have only one or two projects to grade each night. If it is a topic project, where students pick an area of mathematics that interests them and then do independent research, you may want to talk during the first week of school about your expectations and then have a different team turn in their project every week for the entire semester. Design the time frame to preserve your sanity; estimate the number of hours you can afford to spend grading projects and plan the due dates accordingly.

Integration with Other Disciplines

Math is much more meaningful to students when they can see it being used, both in situations that occur in the world around them and in other classes.

Co-Planning

Co-planning your project with a teacher from another department can provide a terrific opportunity for you to help your students make the connection between math and other fields, and sharing the in-class instruction is a smart way for you and your colleague to minimize the amount of class time drawn away from the rest of your curriculum. If you wanted to assign a project with a written component, perhaps the English teacher could use your assignment to cover the yearly unit on research papers. Talk to the science teacher, determine where your curriculums overlap, and plan a project around one of those topics. Have the art teacher spend some time discussing the components of good poster, collage, or mural design with your class. The important issue here is to approach your colleague with the attitude of "How can we both save time and energy, yet make the material we need to cover more meaningful to our students?"

Co-Evaluating

Successful integration of projects into your mathematics classroom includes avoiding being buried in grading at the end. Sharing the assessment load with another teacher is an obvious way to lighten your burden. It can also serve to motivate students—after all, the project is going to affect their grade in more than one class. Splitting the grading is easiest when you each choose a specific component to evaluate. As the math teacher, you would assess the correctness of the data analysis, while the science teacher may read the lab and assign a grade for method and conclusions, and the English teacher might scan it for grammar, spelling errors, style, and overall organization. A powerful assessment technique is for the members of the evaluating team to grade their particular components individually, then meet as a group to look at the project as a whole.

Guest Speakers

Guest speakers need to be planned well in advance. Find speakers from those who work within the school and those who come from the community.

Colleagues

Teachers from other departments make excellent sources for a quick synopsis or a review lesson. Ask the librarian to give the class a guided tour of the school's research facilities, persuade the government instructor to introduce a unit on voting theory with a quick overview of U.S. apportionment techniques, or beg last year's math teacher to drop in and remind your students of that nifty method for factoring trinomials that she showed them last year.

Community Members

Students have a healthy respect for the opinions of successful people from their towns. What better way to convince them of the importance of the math they are studying than to have local businesspeople explain how they use it everyday? Students' families and retired senior citizens make exceptionally eager sources, and having them come to speak to your class invariably enhances your own reputation within the community (no small thing in this age of yearly town-override

votes for school funding—the goodwill you can generate by involving the local citizenry is amazing). It's important to remember that most of these guest speakers have little or no experience in speaking to students. They will appreciate guidelines on the topic, the cognitive level of your class, and the time limits of their presentation. If possible, have them give you a preview of what they plan to say, so you can assess its interest level and the appropriateness of its intellectual content. Make tactful suggestions for modifications if necessary—for instance, if the engineer makes it clear that the plan is to devote 30 minutes to differential equations in your first-year algebra class.

Flexibility Issues

As the teacher, it is up to you to provide an expectations framework for the project. However, within these guidelines there must be flexibility.

Students as Creators

If students are encouraged to follow their creative impulses, you will find that they take on the pride of ownership and consequently work harder to produce a quality product. Consider how you might encourage originality within the structure of the assignment, and specify in which parts of the project students are allowed to set their own parameters. For instance, you might choose the unit topic "projectile motion and the quadratic formula," but let the students design their own catapult system and summarize their results either on a poster or as part of an overhead presentation. Given this freedom, some students will exceed your expectations and spend hours coming up with homemade rockets and elaborate computer-generated graphics to give their project that truly unique stamp.

Meeting Students' Needs

While a project should be designed to expose students to different methods of acquiring knowledge, it should also provide options that allow students to perform to the best of their abilities.

Learning Styles

Students learn in different ways. A visual learner, for example, could have difficulty processing ideas from a guest speaker who doesn't use the overhead or chalkboard, so you may want to lead a class discussion afterwards and summarize points on the board, or take notes to photocopy and hand out later.

Another decision is whether to allow students choice in the way they transmit the knowledge they have gathered. A shy teenager may not do well if forced to give a presentation in front of the entire class, so might instead be allowed to present privately to the teacher or to prepare an audiotape. Concrete, sequential learners may get more from distilling their knowledge into a five-page paper than from juggling all the variables involved in producing a video. Remember, the students are conscious of being graded at the end of the project and sensibly want to choose a presentation format that capitalizes on their strengths.

On the other hand, one of our goals as teachers is to turn out graduates who can function successfully in a variety of situations. Therefore, if you plan to demand one particular mode of presentation so that your class becomes familiar with the skills needed to communicate in that medium, it might be a good idea to allow the students to choose their own partners. Advise them that, instead of rushing to pair up with their best friend, they should ideally search for someone whose strengths and weaknesses complement their own. Have them list their particular strong points and areas of concern, then try to match students who might not otherwise work together.

Special Needs

Be especially sensitive to the fact that students with learning disabilities may need extra support, or modifications in the assignment. Some special-education students have trouble organizing large or long-term projects, and will benefit from subdivided tasks and frequent deadlines. Others, who have always done well in traditional math settings, may have trouble with the reading and research components. A confidential survey will help you to determine whether any of your students have special needs and are apprehensive about meeting your expectations. Many schools have learning-disability coordinators who could work with the students to help them fulfill the goals of the project.

Support During the Project

Students work best, and are most successful at meeting your expectations, when they have ample structure and feedback. You may therefore want to incorporate certain features into your design.

Expectations Form

Give this to your class on the first day you discuss the project. It should contain
- a detailed description of the assignment
- the areas of constraint ("your project must have . . . ")
- the areas of flexibility ("you are free to choose . . . ")
- the aspects to be evaluated, broken down by points or percentages
- the proportion of their semester or quarter grade

Initial-Reaction Form

After presenting the objectives and constraints of the project and answering any questions, ask the students to brainstorm, and write down their ideas for
- possible topics
- possible methods of presentation
- ways to make their project original and entertaining
- preferred partners (if appropriate)
- any fears, concerns, or learning disabilities

Return the papers the next day with written feedback explaining which ideas are acceptable and which are not. You will also want to respond to their concerns, and voice any of your own. This helps the students to attune their thinking with your objectives for the project.

Formal-Proposal Form

After students have received your feedback on their ideas, give them time to do some research. Then ask them to turn in their formal proposals, which should include
- topic
- method of presentation
- outline
- bibliography

You may want to impose additional requirements on the bibliography to ensure that there is adequate research material to support the topic they have chosen. Also, requiring the class to staple their proposals to the initial reaction form will streamline your approval process and avoid any arguments over "but you said I could . . .

Whole-Class Discuss[ion]

About a week bef[ore] mock project to according to the someone has fin to volunteer to pr feedback and corre ually determined a g process with them, co with yours. This is inva their own project may n or that they may have int rectly, while they still have

Conferences

Two to three days before their individual or group meet with you before or after school to give an update on th of their product. During this time they should
- explain which parts are complete and which ones they are still working on
- ask for guidance on issues with which they are having trouble
- use the expectations form to assess their project as it currently stands so that they can troubleshoot any glaring deficiencies while they still have time to repair them

A conference helps keep them from waiting to pull the project together until the night before it is due, minimizing requests for extensions.

Evaluation

Assessment of the project will be a part of your overall grading plan.

Replacement of other Assessments

A well-designed and implemented project should be able to take the place of a unit test. This would free you from having to create, administer, and grade an exam, in a sense reimbursing you for your investment of time and energy in developing the project. To determine whether it is appropriate to replace the exam with the project, find a sample of a test you have given on this unit in years past and list the skills and knowledge assessed. Then be sure your expectations form demands that the students demonstrate skills and that your grading criteria are con o measure whether they actually do. If there ioles whereby the students can meet the goals toject without mastering all of the unit objec may be necessary to give a short quiz to see if ave learned all of the relevant material.

[As]sessment Components

important that students understand the criteria the project from the first day, so that they can ign their projects to meet your expectations. though you will obviously have many mathematical omponents specific to the project you have assigned, here are a few general categories you may also want to consider including.

Social Behavior

If students have been working in cooperative-learning groups, it is often appropriate to base part of their individual grade on
- their effectiveness in working with others (Measure this through in-class observations of specific behavior.)
- their contribution to the team product (The group itself is often the best judge of whether everyone pulled their fair share, or whether one or two people did the majority of the work.)

Research
- Is the background or historical material correct and relevant?
- Is the bibliography varied?

Error Analysis

Have the students list some factors that could make their conclusions or predicted outcomes inaccurate, especially if they generated their own data.

Aesthetic Appeal

Given the importance of presentation skills in the business world that awaits your students, it is crucial that they begin to realize that appearance is often as critical as substance. You will want to make them aware of

- neatness (both in personal and project appearance)

- technical correctness (grammar, punctuation, spelling)

- technology (prepackaged software creates better graphs and pie charts than hand-drawn illustrations)

- oral presentation skills (overhead slides prepared, speaker not overly dependent upon notes)

Creativity and Effort
Did the student work hard to produce an innovative and quality product, or did she merely settle for doing the bare minimum necessary to earn an acceptable grade? Give extra credit for products that evidence original thought and effort above and beyond the expected.

Organizational Checklist

Use this checklist to help you structure your expectations for the project you intend to assign. Each heading refers back to its counterpart in this chapter.

Topic
Presentation format   _____

  _____

Purpose
Cognitive Goals

  List them   _____

  _____

  Does the project meet all these goals?   _____

Background

  List skills and knowledge needed   _____

  _____

  Is a review necessary?   _____

Social Goals

  List them   _____

  _____

  Will you provide feedback?   _____

  Will you provide assessment?   _____

Time Frame
Students' Needs

  Class time allowed   _____ days

  Independent work expected   _____ nights

  Will there be progress checks?   _____

  Late penalty   _____

  Extension policy   _____

Your Needs

  Will you collect projects all at once?   _____

  Number of hours to grade each project   _____

  Total grading hours   _____

Organizational Checklist

Integration with Other Disciplines
Co-Planning

Are you working with another teacher?   _____

You will cover   _____

                                           _____

They will cover   _____

                                           _____

Co-Evaluating

Are you grading with another teacher?   _____

You will assess   _____

                                           _____

They will assess   _____

                                           _____

Guest Speakers

Colleagues?   _____

Community figures?   _____

Person, topic, time allotted   _____

                                           _____

                                           _____

Flexibility Issues
Students as Creators

Which parts encourage originality?   _____

Meeting Students' Needs

Will you allow choice of format?   _____

Will you allow choice of partner?   _____

Modifications for LD students   _____

                                           _____

Support During the Project
Which of the following will you have?

Expectations Form   _____

Initial Reaction Form   _____

Formal Proposal Form   _____

Whole-Class Discussion   _____

Conferences   _____

Organizational Checklist

Evaluation

Replacement of other Assessment

    Will the project replace the test?   _____

    What skills/knowledge are on the test? _____

    _____

    Is mastery of these skills/knowledge

    a) demanded in the expectations?   _____

    b) assessed in the evaluatory criteria?   _____

    If not, how will you measure them? _____

Assessment Components

    Which of the following will you include?
    What weight will it have?

    Social Behavior   _____

    Research   _____

    Error Analysis   _____

    Aesthetic Appeal   _____

    Creativity   _____

    Effort   _____

Part Three

Implementation

Implementation

Now that your project design is complete, it is time to implement it. Projects are most successful when students are both mentally and emotionally prepared for the task that awaits them. So before the students choose their topics and start their research, invest some time in helping them see that this is an opportunity, not a burden.

Motivational Techniques

It is well worth the time to spend ten minutes or so talking to your classes about why this project is important. Students need to hear that mathematics is more than just a regurgitation of skills. It is, in fact, a tool meant to be creatively applied to finding solutions to real-life problems.

This is your chance to get students excited about the responsibilities they are about to take on, so be sure to stress the positive aspects involved in an assignment of this scope. For instance, unlike an ordinary math test, where students are placed in a timed situation to produce an immediate answer to a question dictated by the teacher, the project will give them an opportunity to either follow up on an interest and add to their personal knowledge or to use the mathematics they have been studying in an original and intelligent manner to model the world around them. They will be encouraged to make conjectures, to do research to support or disprove these ideas, to get constructive feedback from peers and professionals, and to revise their theses to reflect new information as they gather it.

The revision process is as critical to a math project as it is to an English paper, and many students will be glad to hear that if they are willing to put in the effort needed to improve their first few drafts, there is no reason why they shouldn't be able to produce an outstanding product worthy of an A, regardless of the grades they have received in math classes in the past.

Point out that since it is equally important to be able to share these ideas with others, a large part of the project will be the communication of this new knowledge to the teacher or to the class. Students who are talented speakers, writers, actors, and artists—many of whom may not have excelled in math in the past—will be delighted to discover that they will have an opportunity to display their strengths.

It's also important to let the family know what's going on. Don't forget, these people have been burned before—the ones who went to the drug store for posterboard at 11 P.M. and then stayed up all night working on the science collage that Johnny forgot to mention until the evening before it was due. By letting them know your expectations and schedule well ahead of the due date, you gain their trust and support for the nontraditional work you are about to assign. The letter on the next page can be modified to suit any type of project.

Dear Family,

The National Council of Teachers of Mathematics has stressed the importance of students using mathematics to model the world around them and gaining familiarity with techniques for presenting the information they have analyzed. In order to broaden your student's mathematical and technological experience, I have assigned a video project involving applications of sine and cosine waves. Your student received the project expectations form in class today, and we will be using the library's on-line search capabilities tomorrow to come up with appropriate topics and sources. Students will be expected to finish the research on their own sometime over the next four days.

The videotape itself is due a week from Friday. If your child does not have access to video equipment at home, he or she is welcome to sign up any time this week before or after school to use the video cameras and tapes we have at school in the audio-visual lab. In addition, I am available Wednesday and Thursday afternoon if your son or daughter is interested in learning to use our computers to create computer-generated graphics. The project is worth approximately 15% of the quarter grade, and all requirements are spelled out on the expectations form.

I look forward to receiving many thought-provoking and innovative projects. If you have any questions or concerns, please feel free to call me at school. I will return your call as soon as possible.

Setting the Expectations

Once you have completed your project design and have decided which presentation formats you will allow, it is critical to immediately express to students your ideas of what constitutes a good project so that they are channeling their energies in an appropriate direction. Hand out your expectations form, which should contain

- a detailed description of the assignment
- the areas of constraint ("your project must have . . .")
- the areas of flexibility ("you are free to choose . . .")
- the aspects to be evaluated, broken down by points or percentages
- the proportion of the semester or quarter grade

Part One contains sample expectations forms for different projects at both the middle-school and high-school levels.

Choosing a Partner or Team

While an important skill we teach is the ability to work effectively with many different types of people, students enjoy projects more and are willing to put more effort into them when they are allowed some input into their choice of partner. In cooperative learning groups of four, it is often an effective strategy to let students pair themselves off, and then for you to choose the two pairs that will constitute each foursome to ensure adequate race, gender, and ability diversity. As on the Partner Request Form (page 52), students should be asked to justify their choice of partner, so that they will have to think about the strengths and weaknesses a particular friend will bring to the team. Although you as the teacher will sometimes need to impose restrictions, remember that students often ultimately benefit from being allowed the freedom to make dubious choices by learning that this freedom of choice carries with it explicit behavioral and workload responsibilities.

Choosing a Topic

Once students know who their partners are going to be, the groups will need about ten minutes of class time to discuss ideas and formats for their projects. Inform the teams that you want them to come up with several topic and format choices, and that you also want some ideas as to how they are going to make their projects especially interesting or original. During this time, every group member should have on their desk a copy of your expectations form to remind them of the specific guidelines. Then split the teams up and have each individual fill out an initial reaction form. A sample form is on page 53.

Collect the initial reaction forms and look them over by group. While most teams will have submitted forms that are virtually identical, it's easy to immediately spot groups that have difficulty working together by the lack of agreement among their ideas and by the concerns they voice. Make a note to set up a conference with any such team to help them devise ways to seek and to build upon the common ideas they share. With all the forms, provide written feedback as to which ideas meet your expectations for the project and which ones will need to be modified, as well as suggestions for improvement. Also, respond to any concerns or fears the students may have, especially if it seems that a student may need extra help or a modification in the assignment because of a learning disability. Time spent voicing your own concerns about a group's planned project and clearing up misinterpretations of the assignment is time well spent—it will minimize students' frustrations later on, and increase their willingness to work hard to meet the standards you have set.

The next day in class, return the Initial Reaction Forms to the students. Because of their confidential nature, explain to the class that this is one of the few things that you do not want them to store in their portfolio folder. Then hand each team one copy of the formal proposal form (page 54).

Partner Request Form

Confidential

Please list your top three choices for partner on this project.

_____

_____

_____

Answer the following questions for your top choice.

                                                                                        YES   NO   NOT SURE

1. Am I confident that my proposed partner will do an equal share of the work?

                                                                                       ____ ____ ____

2. Am I confident that my proposed partner and I have the behavioral maturity to work together without disrupting the class?

                                                                                       ____ ____ ____

3. What strengths does your proposed partner have for making this project better than a project you would do by yourself?

I understand that working with a partner of my choice is a privilege, and that I may have to forfeit this privilege and work alone if my behavior is not appropriate.

_____
(Signature)

Initial Reaction Form

1. Please list your top two favorite ideas for this project.

2. What are your top two preferred methods of presentation (for example, a video or a paper, or a poster or a skit).

3. Why do you feel your first choice for method of presentation is the best way of presenting what you know about this topic?

4. What ideas does your team have for making this project unique?

5. Do you have any fears or concerns about your team, about your ability to fulfill the expectations for this assignment, or about anything else having to do with the project? If so please briefly describe them here. This information will be kept confidential.

Formal Proposal Form

Our topic: _____

Our mode of presentation: _____

Team members and responsibilities:

NAME	ASSIGNMENT
_____	_____
_____	_____
_____	_____
_____	_____

This form is due tomorrow at the end of class. Please attach the following items to this form:

- outline of your project
- bibliography
- note cards or photocopies of materials from your research

After I have looked over your proposal and either approved or denied it, I will return it to you in your portfolio folder.

Managing a Project

Once you have determined the group configurations, a wonderful organizational technique is to start a portfolio for each team. The students will be generating a great deal of paper—research notes, 75 copies of a survey they administered, proposal forms with feedback from you, spreadsheet graphics to use in the presentation in two weeks—and they sometimes tend to misplace things if left to their own devices. Devote one drawer of a filing cabinet to project portfolios, and start each group off with their own labeled hanging folder with several manila folders inside. Tell the students you expect them to leave everything associated with the project that they don't need for homework in the portfolio at the end of the class period. You will find they are much less likely to lose parts of the project or leave critical pieces at home the day they are needed. Furthermore, when it comes time to grade the end result, you have a history of its development that will help you make a more accurate assessment of the amount of effort, research, and revision that went into the final product.

Managing the Classroom

Unlike the more traditional, lecture-oriented math classrooms they are used to, your students will be placed in a very open-ended, interactive environment for the duration of the class time you devote to working on the project. Although most teenagers will appreciate the unusual degree of freedom, and will work to show you that they are capable of appropriate conduct, there are sometimes a few who feel compelled to test the boundaries of your resolve. For this reason, it's a good idea to emphasize your behavioral expectations before the groups sit down together to start their research.

With older students, it is ordinarily enough to remind the class that working cooperatively is a privilege, not a right, and that on-task behavior is a condition for continued permission to work in groups. Students who are incapable of appropriate conduct can be removed from their teams and sent to a desk in the hall to read their math text and complete a worksheet. One day of enduring the drudgery of the book while listening to peers enjoying their alternative-learning experience is enough to convince most students of the importance of proper behavior.

Some teachers will want to clarify expectations in a contract with the student before readmitting him or her to class. The contract should specify consequences in the event those expectations are not met, including the threat of removal from the class for the duration of the project. Remember, this exercise is supposed to be an enjoyable experience—don't let one knucklehead ruin the atmosphere for you and the rest of the class.

Younger students, or students who have had no previous experience in cooperative learning situations, often need more guidance. Modeling appropriate feedback techniques, assigning roles, circulating with a checklist to assess desired interactions that were discussed at the beginning of the period—these are all effective ways to ensure that your students are actively contributing to a positive, on-task learning environment.

Behavior Evaluation Form

Cooperative Project Groups

The group members were:

Person 1: _____

Person 2: _____

Person 3: _____

Person 4: _____

Behaviors (A plus means I saw the desired behavior, a minus means I observed its antithesis, and a zero means I did not see the positive behavior but neither did I observe its unwanted counterpart.)

	Person 1	Person 2	Person 3	Person 4
Respect for teammates (friendly behavior, no put-downs)	_____	_____	_____	_____
On-task behavior (working diligently to complete the assignment, not goofing off)	_____	_____	_____	_____

Based on the behavior I observed, today your group letter grade is _____

Suggested Classroom Resources

If you are unable to take your class to the library, or if you want them to do some advance research before they go, it's often possible to persuade the librarian to stock a cart with books to be kept on reserve in your classroom for a week or so. Try to obtain

- a dictionary of mathematical terms
- a set of encyclopedias
- old issues of magazines like the *Mathematics Teacher*
- student copies of *Encyclopedia of Math Topics and References,* a book that lists hundreds of topics for math projects, with each topic referenced to commonly available books and magazines that students can use to do research. (If you prefer that your students use computers, this book is also available as a database on disk.) By referencing this resource at the start of the project, not only can groups save valuable time in the hunt for information, but they can also ascertain whether there are enough research sources to support their topic and select another area of study if there are not. The topics referenced are listed on pages 71–73.

If you have assigned a project on a specific topic, the librarian may be willing to collect all of the school's books that deal with that subject and add them to your cart. This prevents a few students from checking out all the relevant materials, making them inaccessible to the rest of the class.

Use the time while the students are doing their preliminary research to hold short conferences with the different groups. Based on their responses on the Initial Reaction Forms, you should have a sense of which teams will need additional guidance. Starting with these groups, work your way through the class, dispensing advice and compliments where appropriate. Remember that this is a new experience for many of your students, and that knowing you are accessible and interested in their progress is reassuring and helps to keep them motivated.

Information-Gathering Forms

While some projects are based on fact-finding trips to the library, others depend upon students gathering their data from live sources. Students who have studied some basic statistics in math class may be more interested in addressing a local issue or a perceived need within their own community than they are in

researching a universal mathematical concept. The two main sources of this type of data are informational interviews and statistical surveys.

Interviews

Interviews are good for gathering background information that is not in print or readily available. Students can tap the knowledge of a person who is a recognized expert in a particular field or who has a unique viewpoint. It is often helpful to interview knowledgeable sources before writing a survey, so that the questions asked in the survey will be relevant and up-to-date. It will help the group's effectiveness—and your own credibility within the community—if you ensure that the students have thoughtfully prepared for the interview by requiring them to phone ahead for an appointment and to type up a list of questions they want to ask. You may want to screen this list for content before they leave to conduct the interview. Remind students to thank the person being interviewed for taking the time to talk to them, and suggest that students also write a note thanking the person for their time and expertise.

If the person is amenable, an audiotape or videotape is an excellent way to be certain the responses are being properly recorded. The group should keep a copy of the tape (or its transcription) in their portfolio. Remind students to include in their project bibliography their interview subject's name, address, and position.

Surveys

Students often need help in designing and administering their surveys. Common mistakes include not making the questions clear or specific enough, not designing the responses to be numerically based (making compilation of the results a nightmare), not polling a randomly selected sample of the population in question, or not polling enough people for the results to be statistically significant. It is a good idea to explain to students that writing a survey often involves a lot of revision, and that they should expect to conduct informational interviews and turn in several drafts before receiving approval to proceed with the polling. Remind the group that you would like them to keep the raw data forms in their portfolio once they finish compiling the statistics.

Sample Interview Form

Student Fees

Name of the person being interviewed:

_____

Position of the person being interviewed:

_____

Name of the people doing the interviewing:

_____

How the interview is being recorded:

_____

1. Is it true that, due to severe budget deficits, the administration is considering charging fees (for equipment, supplies, and moderator fees) to participate in many of the school's extracurricular activities next year?

   _____

2. How big of a budget deficit are we facing next year?

   _____

3. What dollar value are you considering for fees to participate in extracurricular activities?

   _____

4. Would the fees be a onetime charge, or would people who participate in multiple activities have to pay several fees?

   _____

5. Which of the following would be eligible for fees?

   _____ varsity or junior varsity sports

   _____ intramural sports

   _____ newspaper/yearbook/radio station

   _____ forensics/debate

   _____ academic clubs (such as math team, science club, or language societies)

   _____ art/music/drama

   _____ Young Republicans/Democrats

   _____ other—please specify: _____

6. Have any of the following fees been considered for raising money?

   _____ fees for bus to away games/meets

   _____ fees to park in the school lot

   _____ fees to use the writing/computer lab

7. Are there any other revenue-generating ideas under consideration? What are they?

   _____

   _____

8. Are there other areas of the school budget that will need to be cut because of the deficit? What are they?

   _____

   _____

9. Thank you for your time and information, do you have any final comments?

   _____

   _____

Sample Survey Form

Student Fees

Due to severe budget deficits, the administration is considering charging a $50 fee (for equipment, supplies, and moderators) to participate in extracurricular activities next year. Please take the time to fill out this anonymous survey, so that the students' opinions can be heard before any final decision is made.

Background Information

1. Your year in school

   _____ freshman

   _____ sophomore

   _____ junior

   _____ senior

2. Activities in which you participate (check all that apply)

   _____ varsity or junior varsity sports

   _____ intramural sports

   _____ newspaper/yearbook/radio station

   _____ forensics/debate

   _____ academic clubs (such as math team, science club, or language societies)

   _____ art/music/drama

   _____ Young Republicans/Democrats

   _____ other, please specify

   _____

Opinions on Fees

Please indicate your preference by circling the response that most closely matches your own. Use this scale.

5   means you strongly agree

4   means you agree

3   means you are neutral to the idea or don't care

2   means you disagree

1   means you strongly disagree

3. I would support charging a fee of $50 to participate in

varsity or junior varsity sports	5 4 3 2 1
intramural sports	5 4 3 2 1
newspaper/yearbook/radio station	5 4 3 2 1
forensics/debate	5 4 3 2 1
academic clubs (such as math team, science club, language societies)	5 4 3 2 1
art/music/drama	5 4 3 2 1
Young Republicans/Democrats	5 4 3 2 1
other, please specify:	5 4 3 2 1

   _____

4. I would support charging a fee to use the following school facilities if the money would be used to fund the extracurricular program.

fees to ride the school bus to away games/meets	5 4 3 2 1
fees to park in the school lot	5 4 3 2 1
fees to use the computer lab	5 4 3 2 1
other, please specify:	5 4 3 2 1

   _____

5. Do you have any other ideas or opinions you want the administration to hear? Please write comments on the back of this form.

Thank you for your input.

Concurrent Activities

While students are conducting their research or compiling the data they will be using in their project, there is often time for you give short demonstrations of relevant topics. These mini-lessons not only introduce the class to skills that will help them produce a superior project, but also provide a welcome break from the research process and help to keep the interest level high.

How to Use Resources

Too often, we assume teenagers are more adept than they actually are at using the materials and technology available to them. Younger students may benefit from a guided tour of the school's library, with aides or volunteers available to assist them in using the card catalog or in conducting an on-line search.

The computer lab is another area of potential trouble, especially because some students have had exposure to different types of software and programming at home while others are virtual technophobes. Whether it's telecommunications software you use or a simple spreadsheet or graphics program, remember that the class may need some practice with the equipment before they are comfortable and ready to turn out presentation-quality materials.

It is often productive to give the class a short assignment. For example, you could give everyone the same data set and ask them to work in pairs to use business software to produce both a bar graph and a pie chart. Students who are familiar with the program in question can act as "experts" and move about the lab offering assistance to teams that are having difficulty. This minimizes student frustration with technology by sending a strong message to the class that you care about them and are dedicated to providing them with the support they need to do a good job on their projects.

Interactive Skills

Role-playing is an excellent way to help students learn effective interview techniques. Conventions that most teachers view as common knowledge may have to be taught: phoning ahead for an appointment, dressing appropriately for the interview, proper preparation with questions and recording device, active listening skills, thanking the source (both at the end of the session as well as with a letter), etc. Having a dress rehearsal before a group goes out to conduct the interview will often identify trouble areas that the team has overlooked.

Similarly, most students tend to focus exclusively on informational content and don't realize the importance of aesthetics when giving a presentation. Discuss with them—or, better yet, show them—the subtle, favorable impression made upon the audience by neatness, both in the presenter's personal appearance and in the professional quality of the graphics used to illustrate a point. Teach them how to effectively use the overhead projector to highlight crucial information, and stress the importance of technical correctness in grammar, punctuation, and spelling in any charts that are used. Remind them that preparation is the key to a persuasive presentation. Speakers should have practiced to the point where they are not overly dependent upon their notes and can answer relevant audience questions without additional research.

Part Four

Assessment

Once your students have finished their research and prepared their final projects for presentation, you need to be ready to assess them. Because students expect math grades to be readily quantifiable, and because they have invested a lot of effort in these projects, it is crucial that they perceive the grading to be fair and consistent.

From Expectations to Assessment

As your think about your grading criteria, start with the Expectations Form you gave to the class on the day you first introduced the project. Take that form and reword the grading criteria on a separate piece of paper. Give a copy to each student a few days before the project is due so that they can go over their final product with a critical eye before turning it in.

Try to make your assessment criteria as clear as possible, so that students will be able to determine ahead of time what sort of grade to expect and can work to improve their project if they aren't satisfied. A few students who didn't put in enough effort may not like their grades, but they shouldn't be surprised by them.

This is also a good time to reflect on any exceptions you may have decided to grant. If you have an extremely shy or reluctant student, you may want to allow them to participate in a nonverbal capacity (running the overhead projector while a partner explains the significance of the graphs, for example). Or you may have authorized a student with slight motor-skill deficiencies to create a video instead of drawing a poster. Regardless of the special need, you should modify your grading-criteria sheet to mirror your changes and present these students with individualized copies ahead of time.

On the next page is a sample expectations form (taken from Part One) for a video project, along with its grading-criteria counterpart. Notice how easy it is to specify the assessment components if the Expectations Form is complete.

Sample Expectations Form

What

The math project on mathematical modeling you will be working on for the next three weeks will be worth one test grade. You must choose a periodic phenomenon encountered in everyday life, gather data generated by this phenomenon, and use the data to create the graph of a sine or cosine wave and derive its equation. Projects in the past have run the gamut from the ordinary (radio waves, sound or music waves, light waves, lasers, and pendulums) to the unusual (Ferris wheels, tsunamis, car engine piston displacements, lawn sprinklers, and better potato chip designs). I encourage you to select a topic that's unique or one that you find personally interesting.

How

You and your partner will produce a videotape. The final production should be edited for clarity, and should be between 10 and 20 minutes long.

Grading Criteria

- Introduction: Give some background or historical information on the topic you have chosen, and explain any technical terms or concepts. There should be clear indications that you have done some research for this section. [10 points]

- Data: Explain how you gathered and measured your data. Display it in tabular form. [10 points]

- Graph: Plot your data, clearly labeling both axes. Smoothly sketch the curve that best fits the data. [10 points]

- Equation Derivation: Use the data on your graph to derive the amplitude, phase shift, translation, and period of your sine or cosine wave. You must clearly explain on tape where each number comes from. [20 points]

- Application: Use your equation to predict future behavior of the phenomenon via two examples. For the first example you should provide a nontrivial value for the independent variable and solve for the dependent variable; for the second example please provide a nontrivial value for the dependent variable and, because the function is periodic, solve for at least three instances of the independent variable. [20 points]

- Error Analysis: Discuss some of the sources of error that may prevent your equation from accurately predicting the actual behavior of the phenomenon. [10 points]

- Technical Correctness: Your project must meet the specified requirements for length, be edited to remove extraneous scenes, and include a bibliography at the end. [10 points]

- Creativity and Effort: Did you put time and thought into selecting an original topic or coming up with an unusual point of view on a traditional idea? Was this project interesting to watch? [10 points]

Extra credit will be given to projects that evidence extreme effort through wonderfully creative visual or auditory effects, or by taking on an unusually difficult project that goes beyond the mathematical scope of what we have done in class.

Sample Grading Criteria

Videotape
High School Trig Class

Your total score is _____ out of 100 points.
Your letter grade is _____

Introduction _____ out of 10 points.
Did you give background or historical information on your topic, and did you explain technical terms or concepts? There should be indications that you did some research for this section, no matter how common your topic.
Comments:

Data _____ out of 10 points
Did you explain how you gathered and measured your data?
Comments:

Graph _____ out of 10 points
Did you correctly plot your data, clearly labeling both axes and smoothly sketching in the curve that best fits the points?
Comments:

Equation Derivation _____ out of 20 points
Did you use the data on your graph to derive the amplitude (5 points), phase shift (5 points), translation (5 points), and period (5 points) of your sine or cosine wave? In order for you to get full credit on this section, you must have clearly explained on tape where each number comes from.
Comments:

Application _____ out of 20 points
Did you use your equation to predict future behavior of the phenomenon with two example? Example 1: did you provide a nontrivial value for the independent variable and solve for the dependent variable? (5 points) Example 2: did you provide a nontrivial value for the dependent variable and solve for three instances of the independent variable? (15 points)
Comments:

Error Analysis _____ out of 10 points
Did you put independent thought into determining and discussing all likely sources of error that prevented your equation from accurately predicting the actual behavior of the phenomenon?
Comments:

Technical Correctness _____ out of 10 points
Did your project meet the specified requirements? Was it of appropriate length, was it edited to remove extraneous scenes, did you participate equally in both the talking and filming, and did it include a bibliography at the end?
Comments:

Creativity and Effort _____ out of 10 points
Did you put time and thought into selecting an original topic or into coming up with an unusual point of view on a traditional idea? Was this project interesting to watch?
Comments:

Involving Students in the Assessment

Students do a much better job meeting your expectations—and feel more invested in and satisfied with the concept of doing projects—if you allow them to participate in the assessment process. One technique that works quite well is to ask a group that finishes their project early to volunteer to be publicly graded. (Sometimes a few extra-credit points are warranted to offset the inherent risks of being critiqued by the class at large; it is also possible to show a videotape of a presentation from the previous year.)

Give everyone in the class a copy of the grading criteria and ask each individual to read the sheet, take notes while the volunteer group presents, and assign a final grade to their project. Then, without indicating what you thought of the presentation, hold a class discussion, asking students to explain and defend the grades they gave based on the criteria you presented. At the end of the period tell them how many points you gave for each part of the project, so that the audience members can see how well they interpreted the stated goals and expectations of the assignment.

Assessing one project together as a class will help many of your students to see flaws or overlooked areas in their own creations while they still have time to fix them. Also, if you are comfortable incorporating students' opinions of their peers' work into a final grade for the projects, let the students know their opinions will affect the grade. Because students are busy evaluating each others' work, you won't need to worry that the audience will get restless and stop paying attention. Students tend to take their assessment responsibilities very seriously, especially if they know you may ask them to defend a particular grade they gave.

If you are uncomfortable performing an entire-class assessment, it is possible to do the same type of activity on a smaller scale by holding team conferences and guiding the groups through a pre-assessment of their own projects a day or two before they are due. As a matter of fact, some teachers prefer to include a self-evaluation component in the group's final grade, especially if individual grades are being given for a team product and it is unclear whether one or two students may have done the bulk of the work.

Even if the students don't assess each others' work, it might be a good idea to hold them responsible for paying attention to any presentations they watch. If the class was allowed to pick their own topics, the audience is usually quite interested in the variety of presentations. Some teachers may merely need to ask their students to pay attention to the presentations; others may want to demand an informational recapitulation at the end of each presentation. This can be accomplished by requiring them to take notes, to write a summary paragraph, or to answer a question on the next quiz.

Exhibiting the Results

Most of your students will have put a great deal of time and effort into their projects. For some of them, this may have resulted in the best grade they have received in math class in a long time. These teenagers will be justifiably proud of their accomplishments, and will rely on you to provide them with some non-boastful way of exhibiting their masterpieces. One of the easiest ways to do this, for physically tangible projects, is to create exhibit boards or shelves and display them either around the school or in public forums like the local community center.

Some projects are difficult to display, especially skits and business presentations. If you are really organized and have tremendous energy, you could hold a math fair where the community is invited to visit the school, view the projects, and hear the students' presentations. This is a bit overwhelming for most educators, so a simpler method is to capture live performances on videotape and ask the creators of especially fine performances if you can add their project to your video library. Students consider this invitation the highest of all possible compliments, and it benefits you to collect samples of high-caliber work to show to future students. Even bringing a camera to school and capturing them in front of the class lets them know you think that what they are doing is important.

Feedback

After all the grades have been assigned, and models or graphics portfolios have been proudly displayed, the final thing you should do is take time to assess the entire process. How'd you do? What were the components that worked really well? What will you need to change for next year? Did you give the students too much freedom? Too little? Keep in mind that projects never come off perfectly the first time—teachers with years of experience still continually fine-tune their Expectation Forms to reflect improved methods and ideas.

A very good source of feedback is your class. As educators, we tend to be too hard on ourselves, always thinking that we could have done it better. Why not ask your students? Chances are they will tell you it was the most enjoyable thing they have ever done in a math class, and they will also gratefully give you some useful ideas as to how you could make the process even better.

After reading the student Feedback Forms, go back to the Organizational Checklist you used to design the project and make notes to yourself about which parts you want to keep and what areas need improvement. Then put it in a safe place to be revised next fall, put your feet up, and give yourself a well-deserved pat on the back for having the guts to try something different in your math class this year. It's teachers like you who are making a difference.

Feedback Form

Dear Students

Now that your projects are complete, please give me your feedback on which parts of the project design worked well and which areas need improvement. Your honest assessment will be very useful as I revise the system to make it better. Thank you for your constructive feedback. I appreciate and value your ideas and opinions.

1. Did you enjoy working on this project, or would you prefer to have covered the material in a more traditional manner using the textbook and having tests and quizzes?

2. Which format (project or test) helps you to understand and learn more math? Why?

3. What things about the project did you really like?

4. What didn't you like?
   How can I change the project design to improve it?

5. Would you like to do another math project in the future?

THANK YOU FOR YOUR HELP!

Appendix

Math Project Topics

Math Project Topics

Abacus

Abundant and Deficient Numbers

Algebra, History of

Algorithmns

Alphametrics or Crypt Arithmetic

Amicable Numbers

Analyzing Data

Anamorphic Art

Ancient Asian Mathematics

Ancient Babylonian Mathematics

Ancient Egyptian Mathematics

Ancient Greek Mathematics

Angle Trisection Problem

Annuities and Interest

Apollonius (255–170 B.C.)

Apportionment and Fair Division

Arabian and Persian Mathematics

Arabic Numeration Systems

Archimedes (c. 287–212 B.C.)

Archimedian Polyhedra

Architecture

Area Formulas and Calculations

Area vs. Perimeter

Aristotle (384–322 B.C.)

Arithmetic Progressions

Art and Geometry

Astronomy and Mathematics

Atoms and Molecules

Averages

Babbage, Charles (1792–1871)

Banneker, Benjamin (1731–1806)

Bernoulli

Billiards and Pool

Binary Number System

Binomial Theorem and Distribution

Birthday Problem

Boolean Algebra

Bridges

Buffon's Needle Problem

Calculating Devices

Calculating Prodigies

Calculation Shortcuts

Calculus

Calendars and Time

Calligraphy and Typography

Cantor, Georg (1845–1918)

Card and Number Tricks

Cartography and Maps

Casting Out Nines

Catenary Curves

Cayley, Arthur (1821–1895)

Celtic Design

Central Tendency Measures

Chaos

Chinese Mathematics

Circuits

Clock Arithmetic

Collecting Data

Combinations and Permutations

Complex Numbers

Compound Polyhedra

Computers and Mathematics

Conflict

Conics

Consumer Math

Continued Fractions

Crypt Arithmetic

Cryptography and Cryptanalysis

Crystallography

Crystallography Patterns

Cube Root Algorithm

Curves

Curve Stitching

Data Analysis

Decimal Fractions

Decision Making

Deduction vs. Induction

Deficient and Abundant Numbers

Density of Numbers

Descartes, René (1596–1650)

Design

Designs from Many Cultures

Determinants

Difference Equations

Dimensions

Dissections

Divisibility Rules

Domes

Drawing Devices

Durer, Albrecht (1471–1528)

Earth Measure

Earthquakes and Logarithms

Egyptian Numeration

Einstein, Albert (1879–1955)

Elections and Social Choices

Ellipses

Eratosthenes of Alexandria (c. 276–c. 194 B.C.)

Escher, M. C. (1898–1972)

Euclid (c. 300 B.C.)

Euclid's *Elements*

Euler, Leonard (1707–1783)

Euler's Formula

Fair Divisions

Famous Mathematicians

Fermat, Pierre de (1601–1665)

Fermat's Last Theorem

Fibonacci, Leonardo de Pisa (c. 1170–c. 1250)

Fibonacci Numbers

Fields and Groups

Figurate Numbers

Finger Reckoning

Finite Differences

Flexagons
Four-Color Problem
Fourth Dimension and
  Higher
Fractals
Fractions, History of
Frieze Patterns
Fuller, Buckminister
  (1895–1983)
Functions

Galilei, Galileo
  (1564–1642)
Galton's Board
  (Quincunx)
Gambling
Game Theory
Gauss, Karl Fredrich
  (1777–1855)
Gears, Ratios, and the
  Bicycle
Genetics
Geodesic Domes
Geometric
  Constructions
Geometric
  Constructions,
  Unusual
Geometric Designs
Geometric Ornaments
Geometric
  Transformations
Geometry, History of
Germain, Sophie
  (1776–1831)
Goldbach's Conjecture
Golden Section
Graphing and Graph
  Theory

Graphing Calculators
Graphing Data
Gravitation
Greek Numeration
  System
Groups and Fields
Growth Models

Handicraft Designs
Harmonic Mean
Harmonic Sequences
Hebrew Numeration
Hexaflexagons
Hindu-Arabic
  Numeration System
History of Math
Hypatia (370–415)
Hyperbolas
Hypercubes

Imaginary Numbers
Indian Numeration
  Systems
Induction
Infinity
Instruments in
  Geometry
Interest and Annuties
Irrational Numbers
Islamic Art
Iteration and Recursion

Kaleidoscopes
Kepler, Johann
  (1571–1630)
Kepler-Poinsot
  Polyhedra
Kites

Knots
Königsberg Bridge
  Problem
Kovalevskaya, Sonya
  (1850–1891)

Lagrange, Joseph
  (1736–1813)
Land Measure
Laplace, Pierre
  (1749–1827)
Latitude and Longitude
Leibnitz, Gottfried
  Wilhelm (1646–1716)
Life, Conway's Game of
Linear Programming
Line Designs
Line Groups
Logarithms
Logic
Logo Computer
  Language
Logos
Longitude and Latitude
Lovelace, Ada Byron
  (1815–1852)

Magic
Magic Squares and
  Cubes
Maps and Cartography
Map Coloring
Mathematical Induction
Mathematics,
  Definition of
Matrices
Mayan Numeration
Measurement

Measures of Central
  Tendency/Averages
Mental Calculating
  Shortcuts
Metric System of
  Measure
Minimal Surfaces
Mira Constructions
Mobius Strips
Models
Modeling and Decision
  Making
Modern Mathematicians
Modular Arithmetic
Moire Patterns
Molecules and Atoms
Monte Carlo Methods
Motion
Music and Mathematics
Mysticism

Napier, John
  (1550–1617)
Napier's Rods
Nature
Navigation
Networks and Circuits
Newton, Isaac
  (1642–1727)
Nines, Casting Out
Nine-Point Circle
Noether, Amalie Emmy
  (1882–1935)
Non-Euclidean
  Geometry
Normal Distribution
  Curve
Numeration Systems
Number Theory